ALEXANDER FLEMING

By Richard Hantula

WORLD ALMANAC® LIBRARY

Please visit our web site at: www.worldalmanaclibrary.com
For a free color catalog describing World Almanac® Library's list
of high-quality books and multimedia programs, call 1-800-848-2928 (USA)
or 1-800-387-3178 (Canada). World Almanac® Library's fax: (414) 332-3567.

Library of Congress Cataloging-in-Publication Data

Hantula, Richard.
    Alexander Fleming / by Richard Hantula.
        p. cm. — (Trailblazers of the modern world)
    Includes bibliographical references and index.
    Summary: Recounts the life story of Alexander Fleming, his study of medicine and bacteriology,
and his discovery of penicillin.
    ISBN 0-8368-5083-1 (lib. bdg.)
    ISBN 0-8368-5243-5 (softcover)
    1. Fleming, Alexander, 1881-1955—Juvenile literature. 2. Bacteriologists—Great Britain—
Biography—Juvenile literature. 3. Penicillin—History—Juvenile literature. [1. Fleming, Alexander,
1881-1955. 2. Scientists. 3. Penicillin—History. 4. Nobel Prizes—Biography.] I. Title. II. Series.
QR31.F5H36   2003
616'.014'092—dc21
[B]   2002033124

First published in 2003 by
World Almanac® Library
330 West Olive Street, Suite 100
Milwaukee, WI 53212 USA

Project manager: Jonny Brown
Project editor: JoAnn Early Macken
Design and page production: Scott M. Krall
Photo research: Brian Boerner
Indexer: Walter Kronenberg

The publisher extends a special thank-you to Pfizer and Anna Gasner for their artwork.

Photo credits: © AHI: 5 middle, 9, 11, 12, 14, 16, 17 top, 18, 24, 25, 26, 28, 29 both, 38, 39 top, 43; © Bettmann/CORBIS:
4, 17 bottom, 27 top, 32 both, 40; Martin Bond/Science Photo Library: 41 top; © Scott Camazine/Photo Researchers: 6
bottom; © Hulton-Deutsch Collection/CORBIS: 42; © Jung/Photo Researchers: 27 bottom; © Michael Nicholson/
CORBIS: 5 top; © Pfizer, Inc.: cover, 21, 35 top, 36, 39 bottom; © St. Mary's Hospital Medical School/Photo Researchers:
41 bottom; St. Mary's Hospital Medical School/Science Photo Library: 6 top; © Scimat/Photo Researchers: 33;
© Wellcome Library, London: 5 bottom, 7, 15, 19, 30, 35 bottom

Printed in the United States of America

1 2 3 4 5 6 7 8 9 07 06 05 04 03

TABLE of CONTENTS

Words that appear in the glossary are printed in **boldface**
type the first time they occur in the text.

CHAPTER 1

A REVOLUTIONARY DISCOVERY

Alexander Fleming's discovery of penicillin opened the way to a new era in medicine.

It may seem hard to believe, but as recently as the early twentieth century, doctors had almost none of the powerful medicines and equipment that we take for granted today. People died from illnesses and infections that can now be easily treated. Surgical operations were much more dangerous in those days. Many of the millions of deaths that occurred during World War I, which lasted from 1914 to 1918, were due to wound infections for which no effective treatment existed. A few treatments were available for some diseases, but they often had unpleasant or even dangerous side effects. If not used carefully, they might do nearly as much harm to the patient as the disease they were meant to relieve. But in the following decades, a breakthrough occurred. A new group of drugs called **antibiotics** was introduced. These drugs not only possessed the miraculous ability to stop many types of harmful **germs** but also, in many cases, posed little danger to the patient. The first and most famous of these antibiotics was **penicillin**. It was discovered in 1928 in a humble mold—a sort of fungus—by the Scottish-born British doctor and scientist Alexander Fleming.

PREPARING FOR THE UNEXPECTED

In the past 150 years, medical science has seen more advances than in all preceding history. In the second half of the nineteenth century, when this remarkable period began, a number of important discoveries were made. For example, in the 1860s, the British surgeon Joseph Lister found that deaths due to infections resulting from surgical operations could be markedly reduced with the help of germ-killing chemicals called **antiseptics**.

The very fact that germs, or tiny **microorganisms** such as **bacteria**, are responsible for many types of disease was another of the major discoveries of this time. This discovery was largely due to research by the French chemist and biologist Louis Pasteur and the German physician and scientist Robert Koch in the 1860s, 1870s, and later. Pasteur, who also developed **vaccines** for several diseases, once made a statement about observation and discovery in science that became famous: "Chance favors only the prepared mind." The discovery of penicillin, which has been called one of the greatest discoveries in the history of medicine, bears out this saying because it was the result of a chance occurrence rather than a concentrated effort. One of the laboratory dishes, or **petri dishes**, in which Fleming grew bacteria for study was accidentally contaminated by a mysterious mold. Fleming noticed that something associated with the mold seemed to be killing the bacteria.

Fleming had a natural knack for observation, but he was also trained through years of study and research to notice occurrences that might be of scientific significance. He was well prepared to make the discovery when the opportunity presented itself. Indeed, he had already earned something of a reputation with his work as a researcher at a London hospital and also at military hospitals in France during World War I. He had already made a major discovery thanks to another chance occurrence and his skill at observation. In 1921, he had discovered a chemical, dubbed "**lysozyme**," that occurred naturally in the body and had the effect of breaking up, or "lysing," certain bacteria.

Joseph Lister (top) founded antiseptic surgery, and Robert Koch (above) and Louis Pasteur (below) showed that many kinds of disease are caused by germs.

A reconstruction of how Fleming's lab looked in 1928, when he discovered penicillin.

INTERNATIONAL EFFORT

Although Fleming was responsible for the initial discovery of penicillin, the efforts of many researchers were needed to purify the substance, test its safety and effectiveness, and develop ways to make it in large amounts. In the early 1940s, a team of researchers at Britain's Oxford University, led by Howard Florey and Ernst Chain, obtained a purified form of penicillin and proved that it possessed lifesaving power. By then, however, World War II was under way. Britain was hard pressed by the war, and the country's pharmaceutical industry was not in a position to tackle production of the new drug. Consequently, a crash effort began in the United States to find ways to manufacture penicillin on a large scale. The Americans succeeded in time to save countless lives during the war.

Molecular structure of synthetic penicillin.

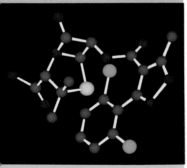

WORLD RECOGNITION

By the time World War II ended in 1945, it was clear to many that a new era was beginning in medicine. Researchers had begun to discover other types of antibiotics, such as streptomycin. The 1945 Nobel Prize in physiology or medicine was awarded to Fleming, penicillin's discoverer, and to Florey and Chain, who played key roles in developing the discovery. Of the three men, Fleming received the lion's share of the honors and public praise that came from countries around the globe. He became one of the most famous persons in the world. This fame was partly due to the fact that Fleming made the initial observation. Partly it was because the story of penicillin's discovery "by accident" caught the public's fancy. And partly it was due to Fleming's unpretentious manner. This highly trained and talented scientist turned out to be not at all an intimidating academic type. People saw a man of below-average height—about 5 feet 6 inches (168 cm)—who was a good-natured and down-to-earth Scot.

Oxford University scientists Howard Florey (back row, far left) and Ernest Chain (back row, second from right) helped turn Fleming's penicillin into a practical drug.

Molds Versus Bacteria

Fleming was not the first to find that certain molds may exert ill effects on bacteria. Old folk beliefs held that some molds could be used as remedies for infection. In fact, the ancient Chinese used moldy soybean curd for infections more than 2,000 years ago. More recently, scientists also occasionally noticed that molds could have antibacterial effects. One example was the famous Joseph Lister, who studied the mold *Penicillium glaucum* in the 1870s. But before Fleming's work with penicillin, the antibacterial substances obtained from molds tended to be very weak. Fleming's mold was an unusually productive rare strain of *Penicillium notatum* that had strong antibacterial activity.

Although Fleming's penicillin was powerful, it did not kill bacteria immediately. Instead, it seemed to act in a delayed manner. Many years after Fleming made his discovery, scientists learned the explanation for this. It turned out that penicillin does not (generally speaking) kill adult bacteria at all. Rather, it works by stopping young, growing bacteria from forming their cell walls.

CHAPTER 2

A SCOTTISH FARM BOY

Alexander Fleming was called Alec by his family and many others; in the lab, however, he was known as Flem. His friends at the Chelsea Arts Club, a favorite haunt when he was not at work, often called him Sandy, presumably because of his fair hair. He was born on August 6, 1881, on his parents' farm about 4 miles (6 km) from the small town of Darvel in the region of southwestern Scotland known as Ayrshire. The farm, set in a hilly area, had sheep along with some cattle, and there were a few dozen acres of land for crops.

The Flemings were used to large families. Alec's father, Hugh Fleming II, was one of nine children. Hugh II married twice. The children of the first marriage included Jane, Hugh III, Thomas, and Mary, along with a child who died while still a baby. In 1874, Hugh's first wife died of tuberculosis, and two years later he married again. With his second wife, Grace, he had four more children: Grace II, John, Alexander, and Robert. Given the meager level of medical care available in those days, it was perhaps not surprising that the family suffered as many deaths as it did. Jane died of smallpox a few years after Alec was born. Alec's father had a stroke in 1887 and then, a year later, also died, at the age of seventy-two. Management of the farm passed to his eldest son, Hugh III.

SCHOOL YEARS

Life on a Scottish farm was not easy. Alec was expected to help out with chores. Still, there was time for fun.

Fishing was one of Alec's favorite pastimes. He also liked to collect bird eggs, which he sold in town for a few pennies. Winters were often marked by severe winds and heavy snows, but winter evenings offered plenty of time for family games. In fact, Alec retained a love for playing games and inventing new ones all his life.

Alec started attending school when he was five. It was a small one-room school about a mile (2 km) from home. On nice days, he often walked there barefoot. In the cold of winter, his mother sometimes gave him two hot potatoes to hold. They not only kept his hands warm on the way to school but guaranteed him a warm meal when he arrived. Later in life, Fleming would recall this tiny school with fondness, saying that he learned a lot, particularly from the opportunity to observe nature and its changes on his daily trek there and back. The opportunity to live on a farm was also invaluable. He said, "We had the farm animals, and the trout in the burns [creeks]. We unconsciously learned a great deal about nature, much of which is missed by a town dweller."

When he was ten, Alec moved on to a larger school at Darvel, again walking there and back home every day, although in extremely bad weather he might stay overnight with a great-aunt in town. One day at the Darvel school, he had an accident that left its mark on him for the rest of his life. Running around a corner, he collided head-on with a shorter boy coming from the opposite direction. The impact perma-

Fleming visits the little school he attended as a child in Scotland.

nently flattened Alec's nose. The accident accounts for the rumors that circulated years later that he was once a prominent boxer.

Alec did well at Darvel, and when he was twelve, his family decided he should continue his schooling rather than stay at home to work on the farm. He entered an academy in the major Ayrshire town of Kilmarnock that offered a wide variety of courses, including science classes. He lived with an aunt in Kilmarnock, going home only on weekends.

THE BIG CITY

Alec spent only about eighteen months at the Kilmarnock Academy. His brother Tom had by then, like many other Scots in search of a living, moved to London. Tom became an oculist, or eye doctor, and set up a home in the big city with his brother John and sister Mary. In time, he invited Alec, then about thirteen and a half, to join them. Several months after Alec's arrival, his brother Robert also came to live with them.

Watchful Eyes

When Fleming was an adult, his eyes and observant gaze were among the main characteristics that struck people who met him. He had big blue eyes, with which he typically stared at a person who was speaking. Some people were confused or flustered by this seemingly cold, blank stare, especially because it was typically accompanied by silence—Fleming was usually a man of extremely few words. There were those who found his stare rude. Women, however, were often said to be fascinated by it. Later in life, after he gained world renown, Fleming became more affable and talkative in public.

Even as a little boy, Fleming drew attention with his remarkable eyes. When he was only eight or nine, he made such an impression on one of his teachers at the tiny school near his family's farm that years later, after he had gained world renown, she wrote a letter to him in which she said she still remembered his "dreamy blue eyes."

Tom enrolled Alec and Robert in the "Commercial Section" of the Regent Street Polytechnic with an eye toward preparing them for a business career. Alec, as usual, did well in his classes. He also took abundant advantage of the opportunity to explore the wonders of London. With its bustling streets, famous buildings, underground railway, and scarcity of greenery, the immense, noisy, dirty city presented quite a contrast to the Ayrshire countryside. As Alec adjusted to the big city, he worked on softening his strong Scottish accent, which sometimes made it hard for Londoners to understand him. Although his accent weakened, it never entirely disappeared.

At the age of sixteen, Alec got a job as a clerk in a shipping office of the American Line, working for pennies an hour. It was at the American Line that he started wearing a bow tie, a habit he would retain the rest of his life. He did not find office work terribly stimulating. About the time he turned twenty, an uncle of his died, leaving him a small inheritance. Alec agreed to a suggestion by Tom that he use it to go to medical school.

Young Fleming poring over his books.

CHAPTER 3

DOCTOR AND SCIENTIST

To enter medical school, Alec had to pass a qualifying examination. During his four years at the shipping office, he later said, he "learned nothing academic." But that did not deter him. He hired a teacher to help him prepare for the exam, which he took in July 1901. The results were phenomenal. He not only passed all the parts of the test but ranked first in the English subjects and tied for first in "General Proficiency." He was now in a position to take his pick from among London's twelve medical schools.

Fleming as a shipping office clerk, around 1900.

According to an account Fleming later wrote, he based his choice on the practical consideration of closeness to his home and his lifelong love of games and sport. He didn't know anything about the three closest schools, except for the fact that one of them, which was associated with St. Mary's Hospital, had a water polo team that he had once played against. So he became one of the eighty students who began their studies at St. Mary's in October 1901. He was to remain at St. Mary's for nearly his entire career.

MEDICAL SCHOOL

Fleming took advantage of the opportunities for extracurricular activities at St. Mary's. In addition to playing water polo, he joined the Medical Society, the Debating Society (apparently his love of competition won out over his habitual silence), and the rifle club as well as the Dramatic Society. His small size made him a natural for female roles.

Meanwhile, he did superbly well in his studies, winning one prize after another without seeming to have to work very hard. "When he read a medical book," said his brother Robert, "he flipped through the pages very rapidly, and groaned out loud when he caught the author making a mistake. There was a great deal of groaning." In addition to possessing a good mind and fine memory, Alec seems to have had a talent for working extremely efficiently, not wasting time on irrelevant things. His chief rival at medical school, C. A. Pannett, later wrote of Alec that "one thing was abundantly clear, that he was a first-rate judge of men, and could foresee what they would do. He never burdened himself with unnecessary work, but would pick out from his textbooks just what he needed, and neglect the rest."

In January 1906, he easily passed the exams qualifying him to work as a doctor. He decided, however, that he wanted to stay in medical school for a more advanced degree—which meant he had to find a job that could provide both the time and money he needed for that goal.

ALMROTH WRIGHT'S LABORATORY

As often seemed to be the case for Fleming, chance decided the next step in his life. Almroth Wright, a star St. Mary's researcher in the study of bacteria, or

Fleming at work in a St. Mary's laboratory.

bacteriology, happened to have an opening for a junior assistant in his laboratory. John Freeman, who worked with Almroth, was passionately devoted to the St. Mary's rifle club. The club's performance in competition would obviously be better if Fleming, who was an excellent shot, stayed at St. Mary's and remained a member. Freeman accordingly persuaded Wright to hire Fleming.

Unlike Fleming, Wright was a tall, big man who was fond of giving his opinions on virtually any subject. He was also a skilled scientist who had developed an important vaccine protecting against typhoid. Wright believed that the best way to fight disease was through vaccination, or immunization with a vaccine, to stimulate the body's defenses —what these days we call the **immune system**—against invading germs. A vaccine for a given disease was typically made from dead germs responsible for causing the disease. "The physician of the future," Wright liked to say, "will be an immunizator." Not long after Fleming joined his staff, Wright reorganized his lab into an "**Inoculation** Department" in expanded quarters. The department not only carried out research but produced vaccines for sale, thereby earning money to support its activities.

Fleming had thoughts of becoming a surgeon—indeed, he took and passed a major exam in 1909 that entitled him to open a surgical practice—but he eventually decided to remain with Wright and pursue a career in bacteriology.

As it happened, 1909 turned out to be one of the most notable years in the history of medicine. The German physician and chemist Paul Ehrlich made a discovery that prepared the way for Fleming's future research on substances like penicillin that fight bacterial **infection** by directly attacking the bacteria rather than by stimulating the body's defenses as a vaccine would. After an extensive search for what he called a "magic bullet" against disease, Ehrlich in 1909 developed a relatively safe chemical that worked against the often fatal disease syphilis. It destroyed the bacteria responsible for the disease without also killing the patient. He called this chemical, which contained arsenic and was the 606th he worked on, **Salvarsan**. This "compound No. 606" was given by injection and carried by the bloodstream into every part of the body. It thus was a "systemic" drug, working on the entire body as opposed to chemicals like the antiseptics doctors were then used to using, which were applied "locally," to just a part of the body. Ehrlich's Salvarsan was the first systemic drug in history to be made by chemical synthesis.

Paul Ehrlich discovered Salvarsan, the first systemic drug made by chemical synthesis.

A quarter century would pass before another chemical would be developed as an effective systemic treatment for disease. The success of Salvarsan, however, drew attention to the possibility that Wright's vaccines might not be the only way to fight infection caused by germs.

Fleming, who had previously developed an improved test for syphilis, was one of the first to try Salvarsan on patients, and he developed methods for

Below: Painting made by Fleming at his country home in Suffolk, north of London.

Artistic Interests and a Playful Side

Ronald Gray, who drew a famous cartoon showing Fleming as a fighter armed with Salvarsan in the war against disease, was one of a number of friends Fleming made among artists in the years before World War I. In that period, he started frequenting the Chelsea Arts Club, where he found good food and pleasant company and could indulge his love of games, ranging from billiards and snooker to croquet to cards. Over the course of his life, he put together a collection of pictures by his artist friends and others. On occasion he himself dabbled in painting, especially watercolors.

A curious hobby he started in the years before World War I remained a pastime of his for years. Many types of bacteria are naturally colored. Fleming would take bacteria capable of showing different colors and arrange the germs in a pattern in a laboratory plate. After a day or so, a picture or design would emerge. "You treat research like a game," Wright told him. But the truth was that Fleming liked to find fun in everything he did, including laboratory research—an attitude that helped him keep his mind open to new ideas and new discoveries.

administering it that reduced the rather painful side effects the drug could cause. He acquired quite a reputation for his skill at curing syphilis. One of his friends, the artist Ronald Gray, drew a cartoon of Fleming in 1910 called "Private '606.'" It shows Fleming standing at attention (and, as was his habit, smoking a cigarette) while holding not a rifle but a syringe containing Salvarsan, or "606."

Fleming as Private "606"—reflecting his work on the use of Salvarsan, or "compound No. 606," in fighting syphilis.

Below: Biologist Selman Waksman, the discoverer of streptomycin.

The Antibiotic Revolution

Synthetic drugs with antibacterial effects came into use before penicillin—notably Salvarsan, in 1909, and the so-called **sulfa drugs**, in the 1930s. But such drugs are often not classed as "antibiotics." The word "antibiotic" is most often applied to substances that not only act against bacteria or other microorganisms but are produced by, or at least are in some way derived from, living organisms. In this sense, penicillin was the first great antibiotic to be discovered, but others soon followed. Tyrothricin, obtained by American **bacteriologist** René Dubos from soil bacteria in 1939, proved to be too toxic for internal use. Streptomycin, discovered in 1944 by American biologist Selman Waksman, also came from soil bacteria. It proved effective against a wide variety of bacteria and for a while served as the chief treatment for tuberculosis. Waksman was the first to apply the name "antibiotic" to such drugs.

Many additional antibiotics were introduced in following decades, some of them representing "semi-synthetic" variations on already existing antibiotics. The search for new antibiotics continues—not only in order to find improvements on existing drugs or to develop treatments for diseases that currently lack them, but also to provide alternatives to antibiotics that become ineffective because bacteria develop resistance to them—a problem that Fleming himself studied and warned about in the speech he gave upon receiving the Nobel Prize.

CHAPTER 4

WORLD WAR I

Fleming in the lab at the British Army hospital in Boulogne, France.

Following the outbreak of World War I in 1914, Wright, Fleming, and other members of the St. Mary's Inoculation Department joined the British Royal Army Medical Corps. They were sent to an Army hospital set up in a former casino in Boulogne, France. A major focus of their research there was the puzzling question of why huge numbers of wounded soldiers died after being brought to military hospitals for treatment. Fleming's brilliant work on this subject earned him a reputation as a leading expert on wound infection.

THE PROBLEM WITH ANTISEPTICS

Surgeons had taken to heart the lesson taught by Joseph Lister in the late nineteenth century that doctors could markedly reduce the risk of infection if they carried out operations without introducing germs into their patients. This meant performing the operations under germ-free, or sterile, conditions. Surgeons and their assistants gradually got into the habit of wearing sterile gowns, caps, and masks. Chemical antiseptics were often used to kill germs in the environment and on surgical instruments. Military doctors, however, were often called upon to work in conditions that were filthy, and it was difficult to keep wounds from becoming infected. In many cases, wounds were already crawling with germs before the patients were even seen by a doctor. Not knowing what else to do, many Army doctors put antiseptic on the wounds in hope of killing the germs, but the antiseptic, according to Fleming and Wright, only made things worse.

Fleming carried out extensive microscope observations, identifying the bacteria involved in wound infections and observing the behavior of the blood cells in the wounds. It was already well known that an important role in the body's defenses against infection was played by certain white blood cells called **phagocytes**, which swallow up foreign invaders such as bacteria. When Fleming studied fresh wounds that had not been treated with antiseptics, he found lots of phagocytes busily at work consuming bacteria. But when he looked at wounds that had been treated with antiseptics, he observed few phagocytes, and most of them were dead or barely alive. He also saw large numbers of living bacteria. Fleming's findings seemed to indicate that treating infected wounds with antiseptics was worse than useless—it was even harmful. This was a message that was difficult to accept for Army surgeons, who firmly believed in antiseptics' power to fight disease by killing germs.

Fleming and his colleagues produced a series of papers explaining what was going wrong. One problem

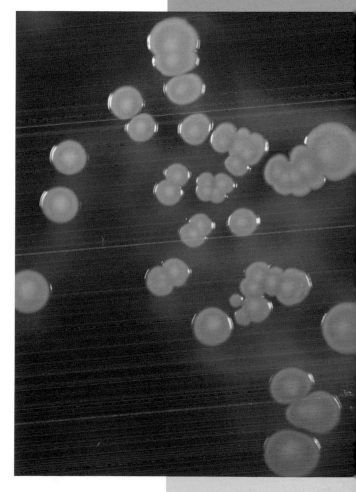

Colonies of staphylococcus bacteria grown in the lab.

was the complicated structure of a wound, particularly a wound made by a bullet or a piece of shrapnel that carried dirt and germs deep into the body. Such a wound has a maze of crevices and winding passageways through which bacteria can find their way into tissues. An antiseptic applied to the top of the wound simply won't penetrate down to the deep bacteria.

Fleming, who was always very handy at building glass apparatus, made a clever "artificial wound" to demonstrate this fact. He heated the walls of a test tube to soften them and then drew out the soft glass to form several hollow spikes. He put an infected liquid in this artificial wound and let it sit for several hours to allow the bacteria to incubate, or grow. He then emptied the tube and filled it with an antiseptic strong enough to kill the germs. After some time passed, he again emptied the tube. This time he filled it with an uninfected liquid, which he allowed to sit for a while. He then observed the original germs growing in this liquid. Apparently the germs had lain hidden in the spikes, where the antiseptic had not penetrated.

But the failure of the antiseptic to reach all the bacteria was not the only problem. Fluid and pus released in the wound by the body tended to carry away the antiseptic, and some antiseptic was also soaked up by dressings and dead tissue.

And there was a third problem—even in areas where it was able to work, an antiseptic tended to do more harm than good. The pus in an infected wound has millions of phagocytes, many of them containing bacteria they have consumed. Fleming showed that treatment with antiseptics killed the phagocytes while allowing the bacteria to live. But that was not all. Most antiseptics, he found, destroyed certain important chemicals, known as **enzymes**, in the blood that played a role in the body's defenses against bacteria.

Some of the difficulties presented by war injuries could be reduced, at least for some types of wounds, by such approaches as removing dirt and dead tissue from the injury, washing the wound with a clean salt solution to encourage the flow of lymph (a body fluid containing phagocytes), and shutting up the wound with sterile dressings. Years would pass, however, before a key need was filled—chemicals, such as antibiotics, that could kill germs without also killing body cells.

DEADLIER THAN WAR

Fleming also contributed to advances in blood transfusion techniques during the war. But his chief research focus continued to be bacterial infection. In 1918, he was assigned to a special hospital that was established at Wimereux, near Boulogne, where he worked on problems of blood poisoning, or septicemia, and the infection known as gas **gangrene**, which was particularly dangerous because it developed very quickly.

Items from the early days of the development of penicillin: a mold colony and handwritten notes made by Fleming.

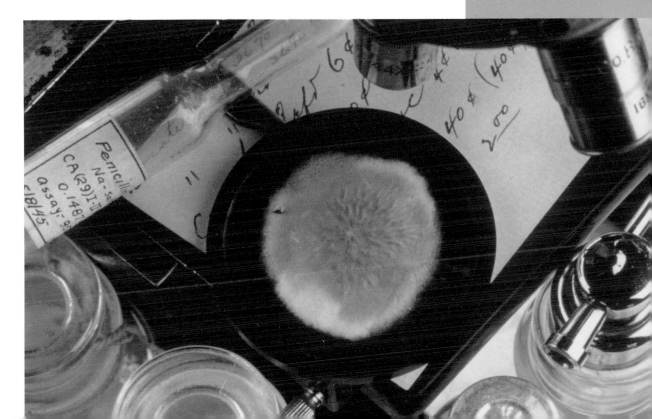

Making Do Under Hard Conditions

Doctors in the early twentieth century had to practice medicine without the advantages of the medical knowledge, equipment, and drugs available to today's physicians. Conditions were especially poor during World War I in the military hospitals hastily set up in the field. These hospitals were crowded, and it was sometimes impossible to observe basic rules governing cleanliness and sanitation. After inspecting several such hospitals early in the war, Sir Alfred Keough, the head of the British Army Medical Services, commented, "In this war, we have found ourselves back among the infections of the Middle Ages."

Fleming and other doctors in the Royal Army Medical Corps (RAMC) did their duty to the best of their abilities, making no exception for holidays. "The picture I have of him," Fleming's assistant later recalled, "is that of a short RAMC officer carrying a tray loaded with pipettes, Plasticine, platinum wire, and a spirit lamp, standing on a cold winter's morning, with ice and snow everywhere, in a tent heated by a brazier, with me carrying out an autopsy on a table, while on another table another corpse lay awaiting its turn! It was Christmas Day, and from each of the bodies Captain Fleming took specimens."

The year 1918 also saw an unprecedentedly deadly influenza epidemic sweep the world. It killed more than twenty million people—a figure that far exceeded the number of deaths linked to World War I. Many scientists at the time believed the flu was caused by a type of bacteria known as Pfeiffer's bacillus, or *B. influenzae*. (It is now called *Haemophilus influenzae*.) Fleming was unable, however, to detect its presence in all patients; in fact, some of the worst cases showed infection with other types of bacteria. He came to believe that bacteria might not be causing the flu but were responsible for secondary infections that helped explain why the epidemic was so deadly. The medical science of that day, unfortunately, was unable to destroy such bacteria lying deep within the body without also harming the body's own tissues. Japanese researchers proved in 1919 that Pfeiffer's bacillus indeed was not the cause of influenza; a **virus**—a tiny microorganism even smaller than a bacterium—was eventually found to be responsible.

TURNING POINTS

The war years showed Fleming death and disease on a scale he had never experienced before, and they sharpened his desire to find a way to heal the deadly infections he saw. "Surrounded by all those infected wounds," he later wrote, "by men who were suffering and dying without our being able to do anything to help them, I was consumed by a desire to discover, after all this struggling and waiting, something which would kill those microbes, something like Salvarsan."

Fleming's personal life also took a new turn in the war period. To the great surprise of his colleagues, he got married while on leave in December 1915. His wife, Sarah ("Sally") McElroy, was a nurse who, together with her twin sister, Elizabeth, ran a private nursing home in London. Sally was about the same age as Fleming, but they were opposites in many ways. He was a short, quiet Scot unaccustomed to emotional display. She was a lively, charming Irishwoman who stood a bit taller than her husband. He was a Protestant. She was Catholic. Their marriage, however, proved a lasting one. Not long after their wedding, by the way, another took place—Elizabeth married Alec's brother John.

CHAPTER 5

"ACCIDENTAL" DISCOVERIES

Fleming's son Robert and wife Sarah.

World War I ended in late 1918, and by January 1919, Fleming was back at work at St. Mary's. In 1921, he and his wife bought a charming country house at the village of Barton Mills in Suffolk, north of London. Here he could once again enjoy the pleasures of nature that he had loved as a child. The couple started spending nearly every weekend at the house, known by the odd name of "The Dhoon." (The origin of the name is unknown.) In 1924, their son, Robert, was born.

Meanwhile, Fleming's professional career was moving ahead. He became assistant director of the Inoculation Department in 1921, and a few years later he was also appointed to the medical school's new chair of bacteriology. Along with his teaching and administrative duties, Fleming continued his research looking for substances that could destroy bacteria without damaging human tissues. It was during the 1920s that he made the two most important discoveries of his career—lysozyme and penicillin. In both discoveries, chance played a large role.

LYSOZYME

In late 1921, Fleming noticed that something interesting had occurred in a petri dish in which he had placed a bit of mucus from his nose a couple of weeks before, when he had a cold. Yellow colonies of bacteria had developed in the dish. This was nothing unusual—the bacteria could easily have come from the air. But what struck Fleming's eye was that there were no bacteria at all near the mucus. He then put some fresh mucus in a sample of the bacteria, which thereupon seemed to be quickly eliminated. Something in the mucus appeared to be destroying the bacteria.

Excited, Fleming thereupon started an intensive series of experiments to see what he could learn about this mysterious chemical substance. Busily collecting samples from laboratory staff and visitors, he found the substance in mucus from other people, including individuals who did not have a cold, and also in other body liquids, or fluids, such as tears, saliva, blood, and pus, and in many tissues of the body. It seemed to exist not only in humans but also in animals and plants, at least the ones tested.

Fleming in this way discovered that ordinary body fluids contain a natural substance with antigerm—or, as we might say today, "antibiotic"—effects. Wright, who

First Major Discovery

Alexander Fleming spoke of the discovery of lysozyme in his Nobel Lecture.

Penicillin was not the first antibiotic I happened to discover. In 1922, I described lysozyme—a powerful antibacterial ferment which had a most extraordinary lytic effect on some bacteria. A thick milky suspension of bacteria could be completely cleared in a few seconds by a fraction of a drop of human tears or egg white. Or if lysozyme-containing material was incorporated in agar filling a ditch cut in an agar plate, and then different microbes were streaked across the plate up to the ditch, it was seen that the growth of some of them would cease at a considerable distance from the gutter.

But unfortunately the microbes which were most strongly acted on by lysozyme were those which do not infect man. My work on lysozyme was continued and later the chemical nature and mode of action was worked out by my collaborators in this Nobel Award— Sir Howard Florey and Dr. Chain. Although lysozyme has not appeared prominently in practical therapeutics it was of great use to me as much the same technique which I had developed for lysozyme was applicable when penicillin appeared in 1928.

Fleming's sketch of a lab dish containing tears in the shape of a T; bacteria added to the dish fail to grow in the vicinity of the T.

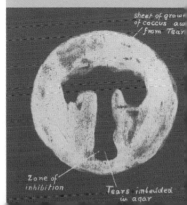

was fond of making up new words from Greek roots, came up with a name for this substance. He called it "lysozyme" because it "lysed," or broke up, the bacteria and was, according to Fleming, an enzyme.

Although lysozyme indeed has antibacterial effects and also, as Fleming showed, leaves phagocytes unharmed, it failed to become the basis for practical medical treatments. Its antibacterial effects are generally quite mild, at least regarding bacteria that cause serious disease in humans. This means that Fleming's discovery was largely a matter of luck. The bacteria that had by chance gotten into the petri dish where he first noticed lyzosyme's effects turned out to be an extremely rare type on which lysozyme acted unusually quickly and strongly. Nonetheless, the discovery of lysozyme was a historic achievement, and the substance remained an important subject of study by researchers for years. For example, Howard Florey and Ernst Chain, who would later play key roles in the development of penicillin as a medical treatment, made key findings about the makeup and action of lysozyme in the 1930s. In the 1960s, lysozyme became the first enzyme for which scientists determined a complete three-dimensional structure.

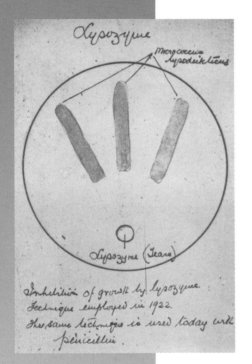

Drawing made by Fleming of an experiment showing that lysozyme hinders bacteria growth.

PENICILLIN

The story of how penicillin was found is similar to the lysozyme story, except that some details are reversed. Lysozyme is a very common substance occurring naturally in the body. Fleming was able to discover it because his laboratory dish containing nasal mucus was accidentally contaminated by a rare type of bacteria that was particularly susceptible to the action of lysozyme. In the case of penicillin, a petri dish containing common bacteria was by chance contaminated with a rare variant, or strain, of mold, presumably from the air, that turned out to be an active producer of a substance with very strong antibacterial effects. Accidental contamination played a part in both discoveries.

Fleming discovered penicillin around September 1928. In one of the petri dishes in which he was growing the type of bacteria known as staphylococci, he noticed a growth of mold. Interestingly, the area around the mold was free of bacteria. He carried out some experiments

Colonies of staphylococcus bacteria grow in the lower part of this lab dish, but not in the upper area near the large blob, Fleming's *Penicillium notatum* mold.

Crystals of the antibiotic penicillin, as seen through a microscope in polarized light

Research Methods

The beginning of Alexander Fleming's paper "On the Antibacterial Action of Cultures of a Penicillium, With Special Reference to Their Use in the Isolation of *B. influenzae*," published in the *British Journal of Experimental Pathology* in 1929:

While working with staphylococcus variants a number of culture-plates were set aside on the laboratory bench and examined from time to time. In the examinations these plates were necessarily exposed to the air and they became contaminated with various micro-organisms. It was noticed that around a large colony of a contaminating mould the staphylococcus colonies became transparent and were obviously undergoing lysis.

Subcultures of this mould were made and experiments conducted with a view to ascertaining something of the properties of the bacteriolytic substance which had evidently been formed in the mould culture and which had diffused into the surrounding medium. It was found that broth in which the mould had been grown at room temperature for one or two weeks had acquired marked inhibitory, bactericidal and bacteriolytic properties to many of the more common pathogenic bacteria.

Colony of Fleming's
Penicillium notatum mold.

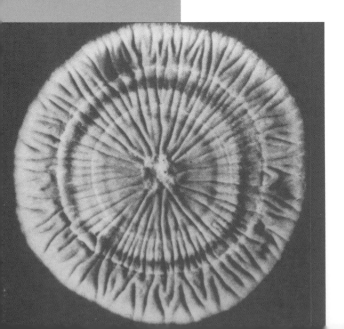

and concluded that that mold contained a substance that had a powerful effect on the staphylococci. This substance was able to stop staphylococci and some other types of bacteria from growing even when he diluted it eight hundred times. Moreover, Fleming found, it was not toxic—that is, it was not harmful to the body. He eventually gave the mold juice, as it was initially dubbed, the more formal name of penicillin, since the mold that produced it belonged to the genus, or type, known as *Penicillium*. (Fungus experts eventually identified the precise species, or subtype, involved as *Penicillium notatum*.)

Fleming reported his discovery in a scientific paper he wrote for the *British Journal of Experimental* **Pathology**, which published it in mid-1929. The paper was titled "On the Antibacterial Action of Cultures of a Penicillium, With Special Reference to Their Use in the Isolation of *B. influenzae*." It opened the way to the antibiotic revolution and thus ranks as one of the most important papers in the history of medicine.

Two close-ups of Fleming's mold as viewed through a microscope.

At the time, however, the paper attracted little attention. More than a decade would pass before penicillin would be developed into an effective drug by Florey, Chain, and colleagues. Although Fleming's paper mentioned penicillin's "possible use in the treatment of bacterial infection," Fleming failed to make its use in medical treatment a focus of his own research.

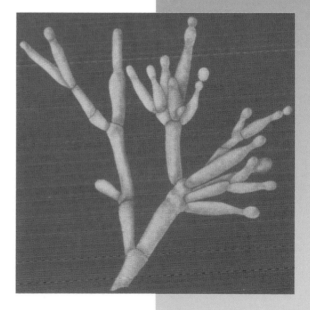

The question of why Fleming left this job to others has fueled much debate among historians. There may have been a number of reasons. For one thing, as the title of the paper suggests, Fleming found his mold juice was useful as a laboratory aid. It helped scientists isolate *B. influenzae* bacteria for study, for example, by

Reprinted from
The British Journal of Experimental Pathology,
1929, Vol. X, p. 226.

ON THE ANTIBACTERIAL ACTION OF CULTURES OF A PENICILLIUM, WITH SPECIAL REFERENCE TO THEIR USE IN THE ISOLATION OF B. INFLUENZAE.

ALEXANDER FLEMING, F.R.C.S.

From the Laboratories of the Inoculation Department, St. Mary's Hospital, London.

Received for publication May 10, 1929.

WHILE working with staphylococcus variants a number of culture-plates were set aside on the laboratory bench and examined from time to time. In the examinations these plates were necessarily exposed to the air and they became contaminated with various micro-organisms. It was noticed that around a large colony of a contaminating mould the staphylococcus colonies became transparent and were obviously undergoing lysis (see Fig. 1).

Subcultures of this mould were made and experiments conducted with a view to ascertaining something of the properties of the bacteriolytic substance which had evidently been formed in the mould culture and which had diffused into the surrounding medium. It was found that broth in which the mould had been grown at room temperature for one or two weeks had acquired marked inhibitory, bactericidal and bacteriolytic properties to many of the more common pathogenic bacteria.

CHARACTERS OF THE MOULD.

The colony appears as a white fluffy mass which rapidly increases in size and after a few days sporulates, the centre becoming dark green and later in old cultures darkens to almost black. In four or five days a bright yellow colour is produced which diffuses into the medium. In certain conditions a reddish colour can be observed in the growth.

In broth the mould grows on the surface as a white fluffy growth, changing in a few days to a dark green felted mass. The broth becomes bright yellow and this yellow pigment is not extracted by CHCl₃. The reaction of the broth becomes markedly alkaline. the pH varying from 8·5 to 9. Acid is produced in three or four days in glucose and saccharose broth. There is no acid production in 7 days in lactose, mannite or dulcite broth.

Growth is slow at 37°C. and is most rapid about 20°C. No growth is observed under anaerobic conditions.

In its morphology this organism is a penicillium and in all its characters it most closely resembles P. rubrum. Bourge (1923) states that he has never found P. rubrum in nature and that it is an " animal de laboratoire." This penicillium is not uncommon in the air of the laboratory.

IS THE ANTIBACTERIAL BODY ELABORATED IN CULTURE BY ALL MOULDS?

A number of other moulds were grown in broth at room temperature and the culture fluids were tested for antibacterial substances at various intervals up to one month. The species examined were: Eidamia viridiascens, Botrytis cinerea, Aspergillus fumigatus, Sporotrichum, Cladosporium, Penicillium, 8 strains. Of these it was found

1

Opening section of Fleming's historic 1929 paper reporting his discovery of penicillin.

getting rid of competing bacteria that were susceptible to penicillin while allowing *B. influenzae*, which are not affected by penicillin, to continue growing. Also, while the mold juice in the form it was available to Fleming may have been perfectly adequate as a lab aid, developing it into a medical treatment would require an enormous

A Matter of Chance

Alexander Fleming described his "accidental" discovery of penicillin in his 1945 Nobel Lecture.

In my first publication I might have claimed that I had come to the conclusion, as a result of serious study of the literature and deep thought, that valuable antibacterial substances were made by moulds and that I set out to investigate the problem. That would have been untrue and I preferred to tell the truth that penicillin started as a chance observation. My only merit is that I did not neglect the observation and that I pursued the subject as a bacteriologist. My publication in 1929 was the starting-point of the work of others who developed penicillin especially in the chemical field.

effort. The mold juice contained many impurities and was unstable. Isolating its active component was an extremely difficult chemical job, beyond the capabilities of Fleming and his assistants, none of whom were professional chemists. Moreover, Fleming's experiments on mold juice's antibacterial effects were not altogether encouraging. His penicillin could take several hours to act on bacteria and seemed to quickly lose its effectiveness if blood was present.

CHAPTER 6

THE ANTIBIOTIC REVOLUTION BEGINS

Fleming's chief contribution to the launching of the antibiotic revolution was his discovery of penicillin. Other scientists, such as Florey and Chain, then played key roles in making penicillin into an effective antibacterial drug—at which point Fleming reentered the penicillin story with a dramatic test on a human patient and with a bid to push the British government to support large-scale production of the drug.

Howard Florey (left) and Ernst Chain (right) pioneered the development of penicillin into an effective drug.

SULFA DRUGS

Florey and Chain's research on penicillin didn't get under way until 1938. It was one of several natural substances with antibacterial effects that they wanted to study. Scientists' interest in substances that could be used to fight bacteria in a systemic way had been fueled by a major discovery made in Germany a few years before—the biochemist Gerhard Domagk had found that prontosil, a kind of dye, was remarkably effective against streptococcal bacteria. Prontosil was the first new synthetic antibacterial drug since Salvarsan. It was the first of several drugs known as sulfonamides, or sulfa drugs for short, which, as a group, proved effective against several types of bacteria. The sulfonamides were hailed as wonder drugs because they were able to save the lives of people sick with previously untreatable diseases such as pneumonia and scarlet fever. The drugs had drawbacks, however. They were relatively toxic, often producing nasty side effects. Moreover, it was found—and this was a problem that would later affect antibiotics as well—that bacteria could develop resistance to the drugs.

Fleming started experimenting with sulfa drugs in 1936, the year after Domagk reported his findings on prontosil. He, along with other scientists, found that sulfa drugs did not kill bacteria but seemed to stop them from growing. The explanation for the drugs' effectiveness, he theorized, was that they weakened the bacteria enough to make it easy for the body's natural defenses to wipe the germs out. In his experiments he found that when a sulfa drug and a vaccine, which boosts the body's defenses, were used together, the results were better than when either was used separately.

Penicillium mold in a close-up view showing the tiny sphere-like spores by which it reproduces

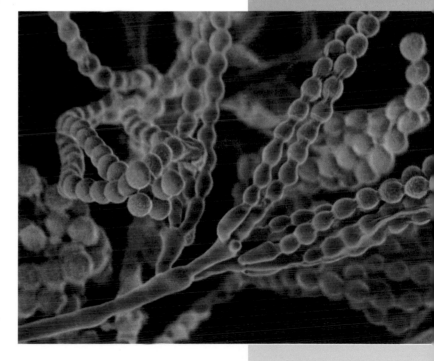

With the outbreak of World War II in 1939, Fleming found himself once again spending a considerable portion of his research time on wound infection. He chose to remain with the hospital in London after the Germans began bombing the city. The Flemings' home was one of the targets hit; luckily, they were away at the time.

Meanwhile, Florey's team at Oxford, which was working with Fleming's strain of the mold *Penicillium notatum*, made progress in purifying penicillin and testing it in animals. Fearing that a German invasion of Britain might be imminent, Florey, Chain, and a couple of others took precautions to keep their promising research, which bore the potential to save countless lives, from

A Concentrated Effort

Alexander Fleming discussed the history of penicillin in his Nobel Lecture.

In 1929, I published the results which I have briefly given to you and suggested that it would be useful for the treatment of infections with sensitive microbes. I referred again to penicillin in one or two publications up to 1936 but few people paid any attention. It was only when some 10 years later after the introduction of sulphonamide *had completely changed the medical mind in regard to chemotherapy of bacterial infections, and after Dubos had shown that a powerful antibacterial agent,* gramicidin*, was produced by certain bacteria that my co-participators in this Nobel Award, Dr. Chain and Sir Howard Florey, took up the investigation. They obtained my strain of* Penicillium notatum *and succeeded in concentrating penicillin with the result that now we have concentrated penicillin which is active beyond the wildest dreams I could possibly have had in those early days.*

being completely lost. They smeared the linings of their coats with spores from the Penicillium mold in hopes that, if the Germans came, at least one of the group might make it safely to another country where the work could be continued.

In August 1940, Florey's team published their findings up to that point in a scientific paper in the medical journal *The Lancet*. This paper, "Penicillin as a Chemotherapeutic Agent," marked the dawn of the antibiotic era. On learning of the researchers' achievements, Fleming went to Oxford to see them in person. "I've come," he said, "to see what you've been doing with my old penicillin." Although they may have been a bit put off by Fleming's sense of ownership, the Oxford group showed him in detail what they had accomplished and gave him a sample of their purified penicillin. Fleming, as was his habit, said very little and left.

Despite the advances made by the Oxford group, the preparation of purified penicillin remained a very difficult process. In 1941, Florey and a colleague went to the United States to promote large-scale production in that country—an effort that ultimately proved enormously successful.

In August of the following year, Fleming used penicillin to produce a dramatic cure in a patient with meningitis. After sulfa drugs failed to eliminate the infection, Fleming, who lacked a supply of purified penicillin of his own, asked Florey for help. Florey sent all he had, and Fleming began to give the patient a series of injections, which produced limited improvement. Fleming discovered that the penicillin was failing to enter the

Technicians at work processing penicillin in a pharmaceutical lab.

An early form of penicillin being administered to a patient through a long, thin tube.

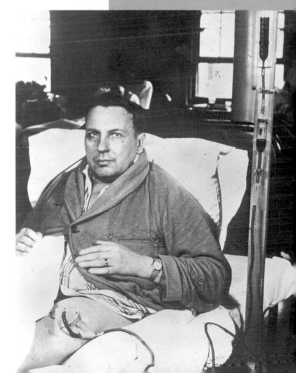

Mass Production

In his Nobel Lecture, Fleming described penicillin production.

I had the opportunity this summer of seeing in America some of the large penicillin factories which have been erected at enormous cost and in which the mould was growing in large tanks aerated and violently agitated. To me it was of especial interest to see how a simple observation made in a hospital bacteriological laboratory in London had eventually developed into a large industry and how what everyone at one time thought was merely one of my toys had by purification become the nearest approach to the ideal substance for curing many of our common infections.

And we are not at the end of the penicillin story. Perhaps we are only just at the beginning. We are in a chemical age and penicillin may be changed by the chemists so that all its disadvantages may be removed and a newer and a better derivative may be produced.

Huge "deep-fermentation" tanks, introduced in the United States, facilitated mass production of penicillin.

spinal fluid, where the infective bacteria remained. Thinking the answer might be to inject penicillin directly into the spinal canal, Fleming telephoned Florey to check his opinion; Florey, however, said he had never tried this seemingly risky approach.

In view of his patient's extremely serious condition, Fleming went ahead and performed the injection. Florey then called back to say he had just tried such an injection on a cat, and the cat had immediately died. Luckily, Fleming's patient recovered.

Impressed by his patient's miraculous recovery, Fleming went to a powerful friend of his, Andrew Duncan, the minister of supply, to appeal for government support for penicillin production. Duncan then spoke to the director general of equipment, Cecil Weir: "Fleming has been talking to me about penicillin. He believes, and so do I, that it offers immense possibilities for the treatment of wounds and of numerous diseases. I want you to do everything you can to organise its production on a great scale." This plea led eventually to the creation of a General Penicillin Committee to oversee commercial production in Britain; Fleming and Florey were among its members.

But it was in the United States that commercial production took off. American scientists and engineers made innovations in manufacturing techniques, notably the use of huge "deep-fermentation" tanks like those used in the brewing industry, to produce penicillin on a grand scale. A worldwide search for a mold even more productive than Fleming's met success in 1943. Mary Hunt, a worker at a Peoria, Illinois, laboratory involved in the project, found an extraordinarily productive strain of the mold *Penicillium crysogenum* on a cantaloupe. Hunt was nicknamed "Moldy Mary" for her dedication to the mold hunt.

CHAPTER 7

WORLD HERO

The public at large heard almost nothing about penicillin until more than a decade after Fleming's 1929 paper. In early May 1941, a few American newspapers reported on a scientific meeting in New Jersey where recent U.S. experiments with penicillin were discussed. Fleming's discovery of penicillin was noted, as was the fact that considerable research was being conducted in England.

The first mention of penicillin in the British press seems to have occurred in late August 1942, in the wake of Fleming's miraculous cure of the patient with meningitis. On August 27, *The Times* of London carried an editorial under the headline "Penicillium." It noted the enormous promise of penicillin and called for speedy development of large-scale production techniques. The editorial gave no names but did indicate that work was being done at Oxford. At the end of the month, the newspaper published a letter from Almroth Wright, who at eighty-one years of age was still officially the head of the St. Mary's Inoculation Department, where Fleming worked. Wright appealed for due credit to be given "to Professor Alexander Fleming of this laboratory. For he is the discoverer of penicillin and was the author also of the original suggestion that this substance might prove

Fleming with his mentor Almroth Wright, head of the St. Mary's Inoculation Department.

to have important applications in medicine." The next day, the paper printed a letter from an Oxford professor drawing attention to the work of Florey's team. These and other newspaper stories touched off a publicity boom, centering primarily on Fleming—a fact that may have been due in part to Florey's reluctance, and Fleming's willingness, to talk to the press. Fleming himself readily admitted Florey and Chain's critical role in developing penicillin into a practical drug. For the press and the public, however, Fleming emerged as the chief hero of the penicillin story, the man responsible for a wonder drug that saved millions of people from disease and death.

Toward the end of World War II, penicillin production surged, saving countless lives at the front. In the United States some twenty companies were making the antibiotic, among them Pfizer and Schenley Laboratories (above, a 1944 Schenley ad).

HONORS GALORE

In March 1943, Fleming was chosen a member of Britain's prestigious Royal Society. Late the same year he was given an "Award of Distinction" from the American Pharmaceutical Manufacturers Association— the first of what was to be a multitude of foreign accolades. In 1944, he was knighted. (Florey was also a recipient of the pharmaceutical association honor and a knighthood, but his awards were paid much less attention by the media.) The Nobel Prize came in 1945. By the time of his death a decade later, Fleming had accumulated more than two dozen honorary degrees and well over one hundred prizes, decorations, and other honors—he was named, for instance, an honorary chief (named "Chief Maker of Great Medicine") of the Kiowa Indian tribe in

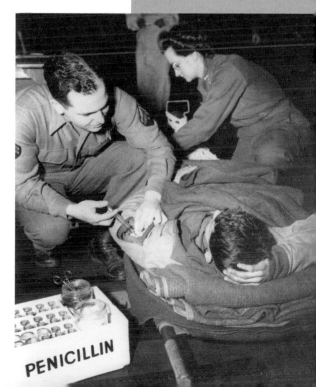

Nobel Prize winners in 1945 included Fleming (second from left), Chain (third from left), and Florey (far right).

1948 and was commissioned a Kentucky colonel in 1954.

Once known as a shy person of few words, Fleming appeared to change after world acclaim befell him. He became more comfortable and outgoing in public. He seemed to enjoy the ceremonial appearances and travel opportunities that came with his penicillin fame. Certainly, people around the world were happy to voice their appreciation. At the Harvard University graduation day ceremonies in 1945, where he received an honorary degree, the 6,000 people present stood and applauded for three minutes when Fleming rose to speak. In 1948, he made a trip to Spain to give several lectures and collect more official honors. The Spaniards' warm welcome was overwhelming. People clapped and cheered when they recognized him on the streets or saw him at the theater or a sports match. He was showered with gifts. At a lecture in Barcelona, a student choir sang a Latin song that had been composed for him, praising him as "the victor of disease and the protector of humanity."

Fleming's discovery of penicillin was depicted in a stained-glass window in a London church.

A Story of Success

Fleming described the effect of penicillin in his Nobel Lecture, December 11, 1945.

I had tested all the chemicals which were used as antibacterial agents and they all behaved in the same way—in some concentration they destroyed leucocytes and allowed bacteria to grow. When I tested penicillin in the same way on staphylococcus it was quite a different story. The crude penicillin would completely inhibit the growth of staphylococci in a dilution of up to 1 in 1,000 when tested in human blood but it had no more toxic effect on the leucocytes than the original culture medium in which the mould had been grown. I also injected it into animals and it had apparently no toxicity. It was the first substance I had ever tested which was more antibacterial than it was antileucocytic and it was this especially which convinced me that some day when it could be concentrated and rendered more stable it would be used for the treatment of infections.

Buildings and institutes were named after him, as was a crater on the Moon and a popular variety of peony.

FINAL YEARS

Fleming is the subject of a stained-glass window in St. James's Church in the area of London where he worked.

In 1946, at the age of eighty-five, Almroth Wright retired as head, or "principal," of what was then called the Institute of Pathology. Fleming, himself already sixty-five, succeeded his longtime mentor. The following year, Wright died, and his old department was given a new name—the Wright-Fleming Institute.

In October 1949, Fleming suffered another loss. His wife, Sareen (as she had for years preferred to be called, instead of Sally), died after a long illness. Her passing deeply affected Fleming. "My life is broken," he told an old friend. It seemed as if he had suddenly aged twenty years. Fleming threw himself into his work at his laboratory, spending evenings at the Chelsea Arts Club and weekends with his brother Robert. He almost never went to The Dhoon—the memories were too painful.

Fleming's world fame and his administrative duties consumed large amounts of his time in his later years, but he continued to work in his lab.

Gradually, however, his gloom eased. In part this was due to the lively energy provided by Amalia Voureka at the lab. Voureka, a young Greek doctor (born in 1912), had joined Fleming's lab with excellent recommendations in 1946. She later recalled how surprised she had been at their first meeting when she found the famous scientist to be a small, kind man, whose "extraordinary

The Fleming Myth

Fleming was a public figure involved in quite technical work whose details were difficult for the public to understand. Inevitably, he became the subject of rumors and stories in the press that were sometimes astoundingly inaccurate. He was amused by such tales and for years collected clippings of them in a sort of scrapbook he called "Fleming Myth." Some he even posted on his department's bulletin board. One example was a story about how Fleming discovered penicillin that appeared in an unnamed British newspaper and was quoted in the *St. Mary's Hospital Gazette* in 1945. It described Fleming as an incredibly busy scientist with so little time to spare that his meals often consisted of cheese sandwiches that he ate in his lab. This unhealthy way of life left him worn out, according to the story, and he developed painful boils. One day, however, he supposedly saw mold on his cheese and suddenly realized that a boil on his neck that had been causing him trouble had vanished. Putting two and two together, so the story went, he concluded the mold had done it—thus was penicillin discovered!

Fleming at work around 1950

eyes seemed to radiate vitality, intelligence, and humanity." In addition to her scientific work, Voureka, who knew several languages, gradually came to serve as Fleming's interpreter during foreigners' visits and as a translator for publication of his lectures in other countries. Fleming began taking her with him to banquets and other ceremonial affairs, and in the summer of 1951 invited her to stay several weeks at The Dhoon.

Having been offered a prestigious job as head of a laboratory in Athens, Voureka left England in late 1951. Fleming missed her deeply, and in October of the following year took advantage of a meeting of the World Medical Association in Athens to go to Greece. Voureka met him at the airport and accompanied him on a monthlong tour of the country. Just before he was to leave for home, Fleming asked her to marry him, and she accepted.

The marriage took place in April 1953 in London. The couple spent their time mainly staying at their home and working there, at The Dhoon, or in travel abroad. In January 1955, Fleming resigned as head of the institute, although he retained his lab. He and his wife planned to leave on a trip to the Middle East in mid-March. But after waking up in good spirits on March 11, Fleming suddenly felt ill and went back to bed. He told his doctor over the telephone that it was nothing urgent. But when Amalia felt his arm, it was cold. He told her, "I'm covered in cold sweat, and I don't know why I've got this pain in my chest." He thought the problem was not with his heart but was puzzled, saying, "It's going down the esophagus to the stomach." But then his head fell forward. He was dead. The cause, according to an autopsy, was a massive heart attack.

In death, Fleming was accorded yet another honor. His ashes were buried in London's St. Paul's Cathedral, alongside those of some of the most illustrious figures in English history.

Fleming in his laboratory at the Wright-Fleming Institute the year before his death

TIMELINE

1881	Alexander Fleming is born on August 6 at his family's farm near Darvel in Ayrshire, Scotland
1895	Moves to London, where he enters the Regent Street Polytechnic school
1897	Gets a job as a shipping clerk with the American Line
1901	Enters St. Mary's Hospital Medical School
1906	Qualifies as a doctor; joins Inoculation Department at St. Mary's, headed by Sir Almroth Wright
1909	Qualifies as a surgeon; begins working with the newly discovered drug for syphilis called compound "606," or Salvarsan
1914	World War I begins; Fleming joins the Royal Army Medical Corps
1915	Marries Sarah ("Sally," later known as "Sareen") McElroy
1921	Discovers lysozyme, an enzyme with antibacterial properties
1924	Fleming's son, Robert, is born
1928	Fleming discovers penicillin
1939	World War II begins
1940	Oxford University researchers led by Howard Florey and Ernst B. Chain succeed in producing relatively pure penicillin and start testing it in animals
1945	Fleming receives the Nobel Prize in physiology or medicine along with Florey and Chain
1946	Fleming named head of Wright-Fleming Institute
1949	Wife, Sareen, dies
1953	Fleming marries Amalia Voureka
1955	Dies on March 11 in London at the age of seventy-three

antibiotic: a drug that kills or weakens germs; often refers particularly to drugs made from substances produced by or derived from life-forms such as molds or bacteria

antiseptics: chemicals that kill germs or stops them from growing

bacteria: certain microorganisms, or tiny life-forms, that consist of one cell

bacteriologist: a scientist who studies bacteria

enzymes: substances in the body that promote a process or chemical reaction

gangrene: death and decay of body tissue, often involving bacterial infection

germs: microorganisms, or tiny life-forms, especially those that cause disease

immune system: the body's natural defenses against germs and foreign substances

infection: the invasion of body tissues by microorganisms

inoculation: the introduction of a substance such as a vaccine into the body in order to help protect against disease

lysozyme: a substance naturally produced by the body that destroys some bacteria

microorganisms: tiny life-forms visible only through a microscope

pathology: the study of the causes of diseases and the changes they produce

penicillin: an antibiotic made by certain molds of the type, or genus, called *Penicillium*

petri dishes: small, shallow laboratory dishes with flat bottoms and loose covers

phagocytes: white blood cells, such as "macrophages," that consume foreign invaders like bacteria

Salvarsan: a drug, discovered in 1909 and originally called compound 606, that was once used to treat syphilis

sulfa drugs: a group of drugs, also called sulfonamides, used to treat certain diseases caused by bacteria

vaccines: substances commonly made from killed or weakened bacteria or viruses that are used to boost the body's defenses against a disease

virus: groups of microorganisms smaller in size than bacteria

TO FIND OUT MORE

BOOKS

Cohen, Daniel. *The Last 100 Years, Medicine.* New York: M. Evans, 1981.

Gottfried, Ted. *Alexander Fleming: Discoverer of Penicillin.* New York: Franklin Watts, 1997.

Jacobs, Francine. *Breakthrough: The True Story of Penicillin.* New York: Dodd, Mead, 1985.

Maurois, André. *The Life of Sir Alexander Fleming: Discoverer of Penicillin.* Translated by Gerard Hopkins. New York: Dutton, 1959.

Otfinoski, Steven. *Alexander Fleming: Conquering Disease with Penicillin.* New York: Facts on File, 1992

INTERNET SITES

BBC: The Modern World—Disease and Its Treatment
http://www.bbc.co.uk/education/medicine/nonint/menus/modtmenu.shtml
Illustrated history of medicine that explores the contributions made by Fleming and other researchers.

History of Biomedicine
http://www.mic.ki.se/History.html
Provides links to a wide variety of online sources on medical history. This site was put together by the library of Sweden's famous Karolinska Institute, which awards the Nobel Prize in medicine.

National Information Program on Antibiotics
http://www.antibiotics-info.org/
A Canadian site focusing primarily on bacteria resistance to antibiotics.

Nobel e-Museum
http://www.nobel.se/medicine/laureates/
Information about the life and work of Fleming and other winners of the Nobel Prize in medicine, including Ernst Boris Chain and Sir Howard Walter Florey.

Penicillin: The First Miracle Drug
http://www.herb.lsa.umich.edu/kidpage/penicillin.htm
An overview of the discovery and use of penicillin, part of the University of Michigan Herbarium's Fun Facts About Fungi site.

About the Author

Richard Hantula has written and edited books and articles on science, medicine, and health for more than two decades. He was the senior U.S. editor for the Macmillan Encyclopedia of Science. Born in Michigan, he has lived in New York City since the late 1970s.